Privacy Technology Implementation Guide

Privacy Office
U.S. Department of Homeland Security
Washington, DC

August 16, 2007

EXECUTIVE SUMMARY

The Privacy Technology Implementation Guide (PTIG) offers assistance to technology managers and developers in understanding privacy protections as they design, build, and deploy operational systems. The guide is offered pursuant to the DHS Chief Privacy Officer's responsibilities under Section 222(1) of the Homeland Security Act of 2002, as amended, to assure that the use of technologies sustain privacy protections related to the use, collection, and disclosure of personally identifiable information. The PTIG is designed to allow each Component the flexibility to adapt privacy considerations to the way that Component does business while retaining a common DHS approach.

The PTIG is not prescriptive. The guide does not mandate the development of any new system requirements. All required privacy protections are governed through the Privacy Office's existing privacy compliance process and the Office of the General Counsel's (OGC) existing legal review process. While this guide makes general recommendations, concerns regarding specific compliance and legal requirements should be addressed directly to the Privacy Office and OGC in order to ensure that all obligations and exemptions from those obligations are properly followed. In addition, this guide does not address the unique nature of technology related to research efforts that are qualitatively different from deployed, operational systems. The Privacy Office and the Science and Technology Directorate are currently discussing a coordinated effort to create a separate PTIG for research.

The PTIG is descriptive. It combines the elements of privacy protection that appear in disparate privacy compliance assessments, documents, and administrative policies and procedures into a single document, contextualized for managers and developers of operational systems. The result is a new orientation guide that provides early awareness of privacy issues and the aspects of systems that can be managed and developed to address privacy issues and streamline the process of complying with existing privacy protection requirements.

This guide is organized into two sections: technology management and technology development. Each section is associated with a particular role within the technology development process and each is designed to address a particular category of privacy protection considerations. The PTIG concludes with a summary checklist presenting the considerations from the body of the guide.

TABLE OF CONTENTS

1. Introduction

The Privacy Technology Implementation Guide (PTIG) is a procedural guide to assist technology managers and developers integrate privacy protections into operational systems that collect, process, or produce personally identifiable information (PII).[1]

Under Section 222(1) of the Homeland Security Act of 2002, as amended, the DHS Chief Privacy Officer is responsible for assuring that the use of technologies sustain privacy protections related to the use, collection, and disclosure of personally identifiable information. The Chief Privacy Officer fulfills this responsibility through a combination of required privacy compliance analysis and documentation and proactive education and collaboration with DHS Components. This guide is a further extension of these ongoing efforts to operationalized privacy protection across the Department.

The PTIG incorporates privacy protection considerations that are currently organized according to the various existing privacy compliance requirements and presents those considerations in the context technologists will encounter them: in the management and development of operational systems.[2]

This guide does not dictate additional mandates for system development. Instead, the PTIG offers a new method of raising awareness regarding what "privacy protection" means in the context of managing and developing operational systems and through that awareness, initiating the process

---

[1] "Personally identifiable information" is any information that permits the identity of an individual to be directly or indirectly inferred, including any other information that is linked or linkable to that individual regardless of whether the individual is a U.S. Citizen, Legal Permanent Resident, or a visitor to the U.S. This definition includes information about DHS employees and contractors.

[2] The PTIG focuses on operational systems. To provide privacy technology guidance for technology research projects, the Privacy Office and the Science & Technology Directorate are discussing a coordinated effort to develop a Privacy Technology Implementation Guide for Research (PTIG/Research) that, as currently envisioned, will provide tailored direction for research program managers and developers to: (1) Identify when a particular research project uses PII; (2) Ensure that nascent technology research and development projects incorporate privacy protections from inception to deployment; (3) Incorporate privacy protections into emerging tools (technology used to build other technology/systems); (4) Integrate privacy protections into the development and life cycle processes that are unique to the field of science and technology research; and (5) Draft all applicable privacy compliance documentation related to particular research projects in a manner that accommodates the unique nature of scientific and technical research.

of privacy compliance earlier in the system development life cycle and more thoroughly across the overall process of deploying systems.[3]

This guide is organized in two sections. The first presents issues related to managing IT systems; and the second presents issues related to systems development.

1. Technology Management. This section describes privacy protection considerations aligned with the administration and orchestration of operational IT systems that can be best addressed by a technology manager reaching across the full range of strategic decisions.

2. Technology Development. This section describes the privacy protection considerations aligned with the specific analysis and functionality of operational IT systems that can be best addressed by a technology developer directly producing the design and development of each portion of the system.

The range of privacy protections relevant to a particular IT system can be addressed through these two perspectives: the overall operation of the system, supported by specific privacy protections built into the system itself.

1.1. Roles

The goal of this guide is to raise awareness of privacy issues for those working directly with technology and to present additional considerations that - if addressed directly and early in system development - can improve the effectiveness and efficiency of complying with privacy protection requirements.

This guide focuses on IT managers and developers in order to identify how, in the context of technology management and operations, privacy protections can be incorporated into IT systems. The PTIG uses two roles to embody these two perspectives: program manager and project developer.[4] The guide uses these labels in a general manner to identify the individuals who are subject matter experts in the management and development of IT systems. As this guide is implemented in practice, these general roles should be matched to specific assignments for

---

[3] With the exception of the specific privacy compliance documents required by the Privacy Office, the PTIG recommendations focus on general privacy issues related to operational IT systems. The PTIG is not a catalog of new documents to produce.

[4] While this guide refers generally to the roles of program manager and project developer, DHS specifically defines the role of "Program Manager" in the DHS Sensitive Systems Policy Directive 4300A as being "responsible for ensuring compliance with applicable Federal laws, directives and Departmental policy governing the security, operation, maintenance and privacy protection of technologies, information and programs under his or her control." In effect, this guide further develops the "privacy protection" responsibilities of the program manager.

individual projects.[5] Distinguishing between the two roles (manager and developer) emphasizes that some of these privacy technology recommendations focus on the overall delivery of technology projects (the program manager)[6] and other recommendations focus on technical features of the systems themselves (the project developer). No distinction is made between responsibilities that should apply to federal employees versus contractor staff.[7]

The responsibility to provide privacy compliance does not lie solely with IT managers and developers. Others across the Department share this responsibility and work closely together to develop and implement privacy protections. Each Component should have an identified privacy point of contact and ideally a dedicated privacy officer to assist with privacy compliance efforts.[8] In addition to the Component level privacy staff, the Department's Privacy Office works closely with both program managers and project developers on an ongoing basis to guide and assist with integrating privacy protections as well as with drafting the required privacy compliance documentation discussed later in this guide (see section 2.4.2).

1.2. Fair Information Practice Principles

The Privacy Office's privacy compliance policies and procedures are based on a set of eight fair information practice principles (FIPPs) that are rooted in the tenets of the Privacy Act and govern the appropriate use of PII. The PTIG draws from these principles as it identifies opportunities for technology managers and developers to integrate privacy protections into the design and deployment of operational systems. While using the PTIG, managers and developers are encouraged to use these underlying principles to guide strategic and tactical decisions that may not be specifically addressed in this guide. As always, the Privacy Office is available to assist in the

---

[5]   The PTIG uses the term "project" and "system" interchangeably with the subtle distinction that a project is the overall effort that produces the system. In effect, however, the recommendations contained in this guide apply to both the project and system level.

[6]   In addition to defining the role of program manager, Directive 4300A also defines the role of Designated Accrediting Authority (DAA) as being ultimately responsible for formally assuming responsibility for managing and accepting project risks. While not specifically addressed in this guide, the DAA could also be identified as the role responsible for managing privacy technology recommendations presented in this guide.

[7]   The Privacy Act specifically states that for purpose of the Act, government contractors are to be considered agency employees, see 5 U.S.C. § 552a(m)(1). As noted above, however, this guide is not prescriptive and all specific compliance issues must be resolved through a close working relationship with the Privacy Office and OGC.

[8]   An example of privacy compliance efforts is implementation of the Privacy Office's Privacy Policy Guidance Memorandum 2007-01 Regarding Collection, Use, Retention, and Dissemination of Information on Non-U.S. Persons, January 19, 2007, applying the Fair Information Practice Principles discussed in section 1.2, to "mixed systems" - systems that use both U.S. person and non-U.S. person data.

pragmatic implementation of these underlying principles as well as the specific recommendations presented in the PTIG. The following is a summary of the FIPPs.

1. Transparency: DHS should be transparent and provide notice to the individual regarding its collection, use, dissemination, and maintenance of PII. Technologies or systems using PII must be described in a SORN and PIA, as appropriate. There should be no system the existence of which is a secret.

2. Individual Participation: DHS should involve the individual in the process of using PII. DHS should, to the extent practical, seek individual consent for the collection, use, dissemination, and maintenance of PII and should provide mechanisms for appropriate access, correction, and redress regarding DHS's use of PII.

3. Purpose Specification: DHS should specifically articulate the authority which permits the collection of PII and specifically articulate the purpose or purposes for which the PII is intended to be used.

4. Data Minimization: DHS should only collect PII that is directly relevant and necessary to accomplish the specified purpose(s) and only retain PII for as long as is necessary to fulfill the specified purpose(s). PII should be disposed of in accordance with DHS records disposition schedules as approved by the National Archives and Records Administration (NARA).

5. Use Limitation: DHS should use PII solely for the purpose(s) specified in the notice. Sharing PII outside the Department should be for a purpose compatible with the purpose for which the PII was collected.

6. Data Quality and Integrity: DHS should, to the extent practical, ensure that PII is accurate, relevant, timely, and complete, within the context of each use of the PII.

7. Security: DHS should protect PII (in all forms) through appropriate security safeguards against risks such as loss, unauthorized access or use, destruction, modification, or unintended or inappropriate disclosure.

8. Accountability and Auditing: DHS should be accountable for complying with these principles, providing training to all employees and contractors who use PII, and should audit the actual use of PII to demonstrate compliance with these principles and all applicable privacy protection requirements.

1.3. Privacy Compliance Lifecycle

The PTIG draws from the various privacy compliance analysis and documentation requirements[9] to offer a single view of privacy considerations for operational IT systems as a whole – presenting the types of privacy protection considerations that should be addressed in general versus specific requirements to be integrated at each stage of development.[10] Just as the PTIG draws from the various discrete privacy compliance requirements, it also draws from the different stages of the privacy compliance life cycle, again, to provide a single view of privacy protection considerations within the context of IT system management and development.

The following summary of the privacy compliance process provides further background regarding the nature of privacy protection requirements. The privacy compliance process flows through four stages:

1. <u>Initial Contact and Coordination</u>. The first stage identifies the project and determines the applicable level of required privacy compliance analysis and documentation. The Privacy Threshold Analysis (PTA) documents the results of this stage. Projects that are identified in the first stage as "privacy sensitive" receive further privacy compliance analysis in the second stage.

2. <u>Collaboration and Development</u>. The second stage integrates the applicable privacy compliance analysis and documentation into the development of the project to ensure that privacy protections are included in the set of business requirements and, ultimately, in the design and deployment of the system. The Privacy Impact Assessment (PIA) and System of Records Notice (SORN) document the results of this stage.

3. <u>Reporting</u>. The third stage reports on the status of privacy compliance analysis and documentation related to each project and the overall DHS system inventory. Departmental reporting to the Office of Management and Budget (OMB) documents the results of this stage.

4. <u>Auditing</u>. The fourth stage analyzes the project's incorporation and performance of all applicable privacy compliance requirements. Specific audit reports document the results of this stage.

The PTIG offers this procedural overview and the specific considerations below to increase awareness of the steps that can be taken to integrate privacy protections into the deployment of operational systems across the Department. The information offered here does not add additional

---

[9] The primary privacy compliance documents are briefly listed in this section and explained in more detail in section 2.4.2, below.

[10] Separately, the Privacy Office is working with the Office of the Chief Information Officer to build specific privacy compliance requirements into the DHS System Development Life Cycle (SDLC).

requirements to the development and operational environments. All privacy compliance requirements are managed through the existing procedures of the Privacy Office and OGC. Hopefully, with the help of this guide, the existing privacy compliance procedures should become more effective and efficient and improve the already productive partnership between the technical, legal, and privacy teams as they work together to satisfy DHS's privacy protection obligations under the Privacy Act of 1974, as amended, the E-Government Act of 2002, and the Homeland Security Act of 2002, as amended. The remainder of this guide presents specific privacy protection considerations relating to IT system management and development.

2. Technology Management

This section of the PTIG offers recommendations regarding how to manage projects in order to assure that the Department's use of technology sustains privacy protections.

2.1. Identify General Descriptions

The first step in developing a privacy protective project is to gather basic information about the project and identify a unique name and a definition that is easily understood and sufficiently unique.[11] This general description will support a quick and thorough understanding of each project's basic purpose and operation and facilitate the alignment of projects and policies.

The program manager should ensure that the following information is identified for each project and is used consistently in all contexts in which the project is discussed, developed, assessed, and documented.[12] Compiling a single authoritative description for the below categories will ensure that all documentation is consistent and all participants remain coordinated and clear regarding each particular project and the full catalog of all projects in the Department.

1. The name of the project. The full name should be the official FISMA Reportable System Name as determined by the Office of the Chief Information Security Officer (OCISO) process for inventory management and change control. If a short name is used, that should be the only short name used. For added clarity, the FISMA ID may be used to supplement a short name. If the project name changes over time, or if the project is divided into sub-projects, the name should be updated across all documentation.

2. The primary point of contact. The primary point of contact is the federal employee primarily responsible for the success of the project. Contact information should include the specific organizational entity (Component, group – down to the most specific organizational unit), title, email address, and phone number.

3. The current status. The current development status of the project should identify whether this is a new development effort or an update to an existing project. The status description should be compliant with the FISMA reportable status for accurate and consistent reporting.

4. The specific purpose of the project. The purpose statement should describe the project's direct contribution to a particular aspect of DHS's mission and should be

---

[11] There are similar requirements for this type of information as part of the FISMA compliance process and as such, the general description would most effectively and efficiently be developed in cooperation with the FISMA reporting requirement.

[12] Each of the required privacy compliance documents (PTA, PIA, and SORN) uses various elements of this general description.

consistent with any documentation related to the Capital Planning and Investment Control (CPIC) process.

5. <u>The data description</u>. The description of the data used in the project should be sufficient to explain what information is needed to accomplish the project's purpose and create the context for any use of PII. Data types should also be defined in the project's FIPS 199 documentation.

6. <u>Personally identifiable information</u>. Personally Identifiable Information is any information that permits the identity of an individual to be directly or indirectly inferred, including any other information that is linked or linkable to that individual regardless of whether the individual is a U.S. Citizen, Legal Permanent Resident, or a visitor to the U.S. This definition includes information about DHS employees and contractors.

   The description of PII should address all PII collected and/or used, the population of individuals related or affected by PII, the role served by the use of PII, the authority to use PII, how the use of PII contributes to achieving the project's purpose, and, briefly, how access to PII will be controlled.

2.1.1. Identify the Project's Full Scope

The program manager should consider the entire scope of the project when gathering the information for the general description and particularly when identifying PII. The scope of concern related to the project should be broad enough to include any secondary or integrated technologies and any up-stream or down-stream data flows (internal or external to DHS) so that the entire life cycle of PII is covered.

2.1.2. Identify the Primary Technologies

Individual types of technology may raise discrete privacy issues. It is important to identify these issues at the front-end of the development cycle so that privacy guidance and requirements can be fully integrated into the design, development, and operation of the system.

The program manager should identify the project's "primary" technologies - those technologies that most accurately characterize the operation and purpose of the project. For example, if one of the project's defining characteristics is the use of biometrics to verify an individual's identity, then one of the primary technologies would be biometrics. If part of the project also involves the wireless transmission of the biometric identifier, then another primary technology would be wireless technology. Identifying the primary technologies improves the efficiency of applying the privacy compliance process to the project. The specific privacy issues triggered by each technology can be identified early in the project's life cycle and those same issues can be properly aligned across multiple projects to ensure that the Department as a whole uses technology in a way that consistently complies with privacy protection requirements.

The following is a partial list of privacy sensitive technologies - technologies that specifically raise privacy issues either through the way the technology processes PII or through the nature of the PII used by the technology.[13]

- Biometrics. Biometric technologies involve the direct use of an individual's physical characteristics and seek to establish the most reliable link between a person and information. The intimate nature and potentially permanent direct association of biometric information with an individual raises privacy concerns regarding risk to the individual from data loss and surveillance (involuntary biometric collection at a distance).

- Geospatial. Geospatial technologies involve the use of geographic information. Since every object and every individual is located somewhere, geospatial technologies can serve as a universal link between all other information, objects, events, and individuals. The ability to associate location with an individual over time along with all other objects and events associated with the same location raises privacy concerns related to tracking and profiling.

- RFID/Wireless. Radio Frequency Identification (RFID) and other wireless technologies involve the transmission of information through the open air. When these technologies are used to transfer PII or are associated in any way with individuals, these technologies raise privacy issues regarding surveillance and involuntary identification. The broadcast nature of the transmission and the association of that data traffic with an individual raises privacy concerns that should be addressed early in the project life cycle.

- Datamining. Datamining technologies generally involve the combination of large volumes of data of various types from many different sources. The potential to connect highly diverse information outside the context of the original collection and to predict characteristics of individuals raises privacy concerns related to data quality and notice. Privacy compliance requirements apply to all uses of PII within a datamining system, which means that the each use of each field of PII should be articulated to facilitate the appropriate level of analysis required to ensure privacy compliance.

2.1.3. Identify the Functional Components

The program manager should ensure that the general description explains the functionally of the system – what the system does to and with the fields of PII. A function-oriented description

---

[13] The DHS Privacy Office will continue to work with OCIO to review the full list of technologies catalogued in the DHS Technology Reference Model (TRM) and further identify potential privacy protection issues that can be associated with particular types of technology.

facilitates a unified understanding of the operation and an assessment of the privacy issues triggered by each function of the system including, at a minimum, how the system will:

- Collect information;

- Process information;

- Report information; and

- Dispose of information.

The program manager should ensure that the entire set of integrated services, technologies, and fields of PII are captured in this description.

2.2. Ensure Controlled Operations

In order to reliably identify and address privacy considerations that arise from the use of PII, it is important to maintain a controlled environment across the entire project life cycle. The program manager should ensure that predictable processes exist upon which to base decisions regarding how PII will be used and confirmation regarding how PII was actually used.

2.2.1. Ensure Itemized Review of PII

The focus of the program manager should extend to the most granular level of PII related to the project. This means that the program manager should ensure the appropriate use of each field of PII regardless of the size of data collection, the variety of the data sources, and the complexity of data use or sharing. The program manager should ensure that all uses of each field of PII by all parties related to the project adhere to the Department's privacy compliance requirements and align with the statements in all related privacy compliance documentation. The dictionaries and diagrams described below can be used to facilitate the thorough analysis of the collection and use of PII and to ensure compliant development and implementation.

- Reconcile Data and Process Dictionaries. The program manager should review both the data dictionary recommended in section 3.3.1 and the process dictionary recommended in section 3.3.2 to confirm that both the PII and the use of the PII comply with DHS's authority. As noted below, both the data and process dictionaries should include a field for the justification for each field of PII in the data dictionary and each processing function of PII in the process dictionary. The program manager should review these dictionaries for completeness and reconcile the justification sections of both the process and data dictionaries to ensure that the purpose of each fits within the scope defined by the authority to collect and use PII.

- Verify PII through the data model. The program manager should review the data model described in section 3.3.3 to ensure that the full scope of PII is properly identified. The data model will show how particular data fields are related and which fields feed or combine with which other fields. Any field that is "linked or linkable" to a field of PII is also PII. The data model should reveal the full scope of PII and may be the most helpful mechanism for the program manager to demonstrate compliance with privacy protection requirements.

- <u>Verify the appropriate use of PII through process flow</u>. The program manager should review the process flow diagram described in section 3.3.4 to verify that all uses of PII remain within the scope of the authority governing DHS's use of each field of PII. Prior to the project's actual use of PII, the program manager should ensure that the PII data life cycle remains within the scope of DHS's authority and is properly described in all required privacy compliance documentation.

2.2.2. Ensure Minimization of PII

The program manager should ensure that all collected PII directly furthers the DHS mission and the specific mission of the Component(s) using the PII. Each field of PII should be reviewed regularly to ensure that it is still needed to accomplish the specific business requirement, the responsibility of the Component, and the mission of the Department. If any PII ceases to be relevant and necessary, the field of PII should no longer be used, removed from the system(s), and disposed of according to the policies and procedures established by OCISO and the Privacy Office.

2.2.3. Ensure Use Limitation of PII

The program manager should ensure that all uses of PII advance the purpose(s) specified in the notice given at the time of the original collection. The program manager should focus on three layers of use:

- <u>Direct Use</u>. The program manager should ensure that the project always uses PII in the same way described in the current privacy compliance documentation (PTAs, PIAs, SORNs).

- <u>Secondary Internal</u> Use. The program manager should ensure that each Component interested in a field of PII has a need to know the field of PII in the performance of its official duties.

- <u>Secondary External Use</u>. The program manager should ensure that any external sharing of PII be conditioned upon the requirements that the recipient only use the PII in a manner that is compatible with the purpose for which the PII was originally collected and that all recipients meet the same standards of data quality, integrity, and protection set by DHS. These arrangements should be memorialized in formal agreements or understandings between the parties and recorded, as applicable, in public privacy compliance documentation. The program manager should specifically identify external sharing of PII across international borders as there may be existing international agreements that impact the use of PII. [14]

The program manager is responsible for ensuring that all PII is accurate, relevant, timely, and complete, within each layer of use.

2.2.4. Ensure Operation of a Structured System Development Life Cycle

The program manager should ensure that the project is managed through a structured process to ensure that all uses of PII are identified and that all privacy considerations are addressed within the context of all other related business requirements. The result should be an alignment between the substantive development of the technology, the administrative thresholds for project approval, and compliance with privacy protection requirements.[15]

2.2.5. Ensure Operation of Formal Change Management Process

The program manager should ensure that the project adheres to a structured process governing changes to the project. Just as with the structured development life cycle, a structured change management process supports compliance with privacy protection requirements by ensuring that all uses of PII are articulated and that all privacy protection considerations continue to be met as the system changes over time.

In addition to the value a structured change management process delivers to the integration of privacy protections as a whole, the process can also be used to ensure all changes in use of PII are reflected in updated required privacy compliance documentation.

2.2.6. Ensure Alignment of Business Drivers & Measure of Success

In order to demonstrate that the use of PII is authorized, the program manager should be able to articulate the line of authority permitting the collection of each field of PII from the Department's enabling authority through the Component's responsibility to the specific authorization of the program that would use the field of PII. The ability to articulate this line of authority will provide the context for both the analysis of privacy issues and the required privacy compliance documentation.

---

[14] As noted above, specific analysis, agreements, and documentation must be coordinated with the Privacy Office and OGC to ensure that all obligations and exemptions are properly addressed.

[15] The DHS Privacy Office is currently working with Office of the Chief Information Officer (OCIO) to integrate specific privacy compliance requirements into the DHS System Development Life Cycle (SDLC).

2.3. Ensure Individual Participation

The program manager should ensure that the individual is as involved as is appropriate regarding the collection and use of PII. As a general matter, direct involvement of the individual fosters transparency into and trust of DHS's operations and improves the accuracy and thus usefulness of the information DHS relies upon to fulfill its mission. There may be situations in which it may be inappropriate, given the nature of interest in the information or the individuals involved, to involve the individual directly. In these situations, the program manger should ensure best efforts are used to maintain data quality standards.

The program manager should ensure mechanisms exists to provide the following capabilities to individuals. The mechanisms can be provided directly through the project or indirectly through a separate administrative process. This capability should be treated as a business requirement and be designed and developed along with other functional requirements.[16]

2.3.1. Ensure PII is Collected Directly from the Individual

The program manager should ensure the project collects information directly from the individual, where appropriate. As a general matter, direct collection increases the likelihood that the information is accurate and that the individual understands how and why the PII will be used. As mentioned above, there are certain conditions under which direct collection of PII may produce information that is less reliable (e.g., in law enforcement or intelligence contexts). The program manager should use the business requirement process to identify whether direct data collection will increase or decrease data quality. If it is impractical or inappropriate to collect PII directly from the individual, the program manager should be able to describe the circumstances that prevent direct collection and use all available mechanisms to ensure the highest level of data quality.

2.3.2. Ensure Individual is Provided Access

The program manager should ensure that a mechanism exists for individuals to request access to PII that relates to them and receive a copy of the PII in a format they can understand. The program manager should also be aware of the way this opportunity for individual access is described in the relevant privacy compliance documentation and oversee any methods developed or deployed by the project developer.

---

[16] Although the PTIG draws from the Privacy Act, the E-Government Act, the Homeland Security Act, and the Privacy Office's policies and procedures, the PTIG's recommendations are not themselves requirements. Program managers should use the PTIG as a general guide to raise awareness and identify privacy protections that could be used to enhance the privacy stature of IT systems. Program managers must consult with the Privacy Office and OGC to identify legal and privacy compliance requirements and any exemptions that may apply in particular circumstances to particular programs or systems.

2.3.3. Ensure Individual is Provided an Opportunity to Correct

The program manager should ensure that a mechanism exists to enable the individual to request corrections to PII. The process of responding and reviewing these requests for correction follows a scheduled process that may involve judicial review of the Department's decision. Of particular note to the program manager is the possible requirement that a written statement from the individual be sent to all recipients of the PII. The project should be able to support this level of follow-up distribution including all the back-end identification and tracking of both the PII disclosure and recipients.[17]

2.3.4. Ensure Individual is Provided Notice of Disclosure

To the extent it is appropriate, the program manager should ensure that a mechanism exists to inform the individual of PII that is disclosed. The program manager should ensure that the system is designed to support a reporting of the date, nature, and purpose of the disclosure of PII and the contact information for the recipient of the PII. This type of notice is specific to the individual and is generally provided in response to an individual's specific inquiry. In contrast, public notice goes to the general transparency requirement to publicize the existence of systems using PII and is implemented through required privacy compliance documentation.

2.3.5. Ensure Individual is Provided a Redress Mechanism

The program manager should ensure the availability of a redress mechanism for individuals to challenge results of the access and correction activities described above. An existing example of a redress program is the DHS Traveler Redress Inquiry Program (DHS TRIP) available on the web: www.dhs.gov/trip. DHS TRIP provides a single resource for individuals with questions or who are seeking resolution regarding the screening process in transportation hubs or during border crossings.

If a redress mechanism is not provided, the program manager should gather information sufficient to explain redress opportunities that do exist or the justification and authority for not providing a redress mechanism. The program manager should coordinate with the applicable Component FOIA Office or the DHS FOIA Office via phone, 703-235-0790 or email, foia@hq.dhs.gov. More information about departmental disclosure and FOIA is also available via the DHS FOIA website, www.dhs.gov/foia.

---

[17] For more details about this process and the potential impacts on the management of projects and systems, the program manager should coordinate with the Component Privacy Officer, point of contact, the Privacy Office, or OGC.

2.4. Ensure Coordination of All Documentation

The program manager should coordinate all documentation related to the project to ensure that the collection and use of PII is presented consistently and comprehensively. The program manager should also ensure that all required privacy compliance documentation is completed prior to loading or using PII

2.4.1. Ensure Alignment of All Documentation

The program manager should oversee the alignment of all project documentation such that statements related to the use of PII are accurate, comprehensive, and consistent. Project documentation may include the following:

- Technical documentation including requirements documents, data schemes and dictionaries, presentations, and materials prepared for the DHS technology investment review process;

- Legal documents, including: contracts, agreements (including international agreements), and/or memoranda of understanding or agreements;

- Budget documents and records, including: OMB 300s and OMB 53s;

- Public documents, including: public media reports, talking points, and testimony; and

- Privacy compliance documentation, including: PTAs, PIAs, and SORNs.

All required documentation, particularly all applicable privacy compliance documentation, should be finalized before any other materials are circulated to the public. These public materials, including press releases and presentations, should include the results of the privacy compliance analysis and documentation applicable to the stage of the project's development.

2.4.2. Ensure Completion of Privacy Compliance Documentation

Privacy compliance documentation serves three functions: It satisfies legal obligations, documents the results of the privacy analysis, and provides the public with notice and explanation regarding PII the Department collects and uses.

The program manager should oversee the timely completion of all required privacy compliance documentation related to the project. All privacy compliance documentation must be finalized, approved by the Chief Privacy Officer, and, with limited exceptions, publicly published before any PII is loaded or used. The program manager should ensure that all privacy compliance documentation is substantially completed during the requirements gathering and design phase of the project in order to ensure that privacy compliance requirements are incorporated into the development of the project to avoid potentially costly redevelopment. This requirement applies to both new and substantially changed systems.

The DHS Privacy Office primarily uses three standardized compliance documents.[18] The following is a summary of the nature and workflow of the standardized compliance documents. The templates for these documents and accompanying guidance are available from the Privacy Office.

2.4.2.1. Privacy Threshold Analysis (PTA)

The PTA is an administrative form created by the Privacy Office to efficiently and effectively identify the use of PII and the need for further privacy compliance analysis. The PTA focuses on three areas of inquiry:

- The purpose and status of the project;

- Potential connections with individuals including the use of PII – any use of social security numbers must be specifically identified; and

- The security classifications through the Certification and Accreditation (C&A) process: Confidentiality, Integrity, and Availability.

The program manager should ensure that the PTA is completed and sent to the Privacy Office early in the project life cycle. If SSNs are to be used, the PTA should specifically identify the justification and authority for using SSNs. Upon receipt of the PTA, the Privacy Office determines the applicability of other privacy compliance requirements including the PIA and SORN. The PTA is complete when the Privacy Office validates it and sends the final copy back to the identified point of contact.

2.4.2.2. Privacy Impact Assessment (PIA)

The PIA is a longer form and is generally required for all projects that use PII. The PIA is an assessment document required by the E-Government Act of 2002 and in support of the Department's privacy protection requirements under Homeland Security Act of 2002, as amended. The PIA must be completed, finalized, and approved by the Chief Privacy Officer before PII is loaded or used. The PIA focuses on the following areas of inquiry:

- Information Collection;

- Information Use;

- Information Retention;

- Information Sharing (internal and external);

- Notice;

---

[18] Additional methods may also be used to measure and mitigate potential privacy risks.

- Individual Access, Redress, and Correction;

- Security; and

- Technology.

The program manager should ensure that the PIA drafting process begins as soon as possible after the validation of the PTA. PIAs are drafted through an iterative process involving the program manager, Component Privacy Office, the Privacy Office, and any other DHS representatives with interests or insights into the project. Once the PIA is substantially complete, the Privacy Office will assess the applicability of the SORN requirement. The PIA is complete when the Chief Privacy Officer signs it. As a general matter, all finalized PIAs are generally published on the Privacy Office's website, www.dhs.gov/privacy.

2.4.2.3.   System of Records Notice (SORN)

The SORN is a legal document required by the Privacy Act of 1974, as amended, and applies to a subset of uses of PII. Unlike PIAs, SORNs focus on the nature and use of information not the technology. This means a SORN could be required for paper records as well as electronic systems. A SORN is generally required when PII is used to retrieve other associated information. For example, if a project has a collection of names, that collection would be considered PII but would not in and of itself trigger the SORN requirement. If, on the other hand, the project involved retrieving records through a search of those names, the SORN requirement would be triggered.[19]

The SORN focuses on the following areas of inquiry:

- The name and physical location of PII used by the technology;

- The nature and source of PII;

- The population of individuals related to PII;

- The sharing of PII outside the Department;

- The policies and practices governing the management of PII, and

- The procedures for individuals to know PII related to them exists and the process to access and request corrections to that PII.

The program manager should ensure that the SORN is completed as soon as possible as PII may not be loaded or used until the SORN is published in the Federal Register – which involves OGC

---

[19] Program managers should coordinate with the Privacy Office for more details regarding information that constitutes a "record" and a "system of records" according to the Privacy Act of 1974, as amended.

and OMB clearance. In addition, there may be requirements to allow for public comment regarding certain aspects of the SORN that would further extend the processing time. The SORN is complete when the Chief Privacy Officer approves it and publishes it in the Federal Register.

2.4.2.4.   Overlapping Documentation

There is a one-to-one relationship between systems and PTAs.[20] There could be a one-to-one or a one-to-many relationship between systems and PIAs as multiple systems may use the same type of PII in the same way for the same purpose.[21] There could also be a one-to-one or a one-to-many relationship between PIAs and SORNs as multiple uses of technology (described in multiple PIAs) could involve the same use of PII (described in a single SORN). Program managers should coordinate closely with the Privacy Office to best understand how a particular project should be reflected in the required privacy compliance documentation and how changes in one system (that could be managed by a different Component) could affect the documentation for many other systems.

2.5. Ensure Training is Provided

The program manager should oversee the development and completion of training by all employees and contractors related to a project that uses PII. The training should provide education regarding how the system uses PII, how the system manages and protects PII, and the safeguards and reporting requirements for working with PII within the Department. The details of this training should be coordinated with the Privacy Office.

2.6. Ensure Use of PII is Audited

The program manager should oversee the design, development, management, operation, and logging of PII use related to the project to ensure that all uses of PII can be audited for compliance with privacy protection requirements.

---

[20] This is reinforced through the TAFISMA tracking system managed by OCISO that requires a validated PTA as part of the C&A process.

[21] This also means there could be a one-to-one or a one-to-many relationship between PTAs and PIAs as multiple PTAs could refer to the same use of the same PII to achieve the same purpose.

3. Technology Development

This section of the PTIG offers recommendations regarding how to design and deploy systems in order to assure that the Department's use of technology sustains privacy protections.

3.1. Minimize Collection of PII

The project developer should design and develop the system to use only the minimum amount of PII necessary to accomplish the system's purpose.

The test for necessity is whether the project's purpose could still be served without the field of PII. If the purpose can be achieved without the field of PII, then the PII is not necessary and should not be used. The project developer should ask this question of each field of PII throughout the design, development, and operational stages as well as during any formal change management process.

The project developer must avoid the collection of social security numbers (SSNs) and only collect SSNs pursuant to a specific legal requirement or other authorized purpose and then, only with the specific approval of the Privacy Office. If the project developer identifies the need for a unique personal identifier, the project developer should create a system-specific identifier instead of using the SSN. The project developer should identify and record the specific authorization as part of the requirements gathering process and monitor all uses of SSNs related to the project to assure the actual collection and use is limited to the scope of the authorization.

3.2. Limit Use of PII

The project developer should develop the project such that each field of PII is only used in ways that are required to accomplish the project's purpose.

The test for necessity is whether the project's purpose will still be served without the particular use of PII. If the purpose can be achieved without the specific use of PII, then the particular use of PII is not necessary and should either not be developed or, if it already exists, should be removed. The project developer should ask this question of each particular use of PII during all stages of the project life cycle.

Required system functions that use PII should be identified during the requirements gathering process, during the overall operation of the system, and specifically during any subsequent change management process. The granularity of review for required uses of PII extends to individual screen functions such as view, print, and copy. The project developer should review each process associated with each field of PII and determine whether each specific use directly advances the project's purpose. For example, if the ability for a particular user to view a particular field of PII is not required to accomplish the project's purpose then the project developer should not develop that specific view function. This minimization requirement parallels the "minimize collection" requirement in section 3.1.

3.2.1. Internal Sharing of PII

If fields of PII will be shared internally (inside DHS), the project developer should implement a mechanism to record that each recipient of PII "has an authorized purpose for accessing the information in the performance of his or her duties, possesses the requisite security clearance, and assures adequate safeguarding and protection of the information."[22] Recording satisfaction of this requirement for each field of PII need not be built into the system as it may be easier to implement this process administratively through the requirements gathering process. The project developer should integrate compliance with this requirement with the data and process dictionaries recommended in section 3.3.

A specific process should be identified for each internal sharing of each field of PII and the justification section of the data and process dictionary of both the sending and receiving system should be sufficiently detailed to demonstrate compliance with the "authorized purpose" requirement.

Project developers should also design and deploy systems in compliance with the DHS Management Directive 4300.1 that may require documentation of internal sharing in an Interconnection Security Agreement (ISA), see DHS 4300A Sensitive Systems Handbook, Section 5.4.3 ("Network Connectivity").

3.2.2. External Sharing and Disclosure of PII

If fields of PII will be shared externally (outside DHS), the project developer should implement a mechanism to record that each remote use of each field of PII is compatible with the purpose of the original collection of that field of PII. As with internal sharing, the compatibility mechanism may be administrative (not necessarily built into the functionality of the system) and should be engaged during the initial design stage throughout the project's life cycle.

Similar to the recommendation regarding internal sharing, a separate system process should exist for each external sharing and the justification section of the data and process dictionaries of the system should be sufficiently detailed to demonstrate compliance with the purpose compatibility requirement.

Project developers should implement a mechanism to support the ability to respond to individual requests regarding the details of external sharing of PII (see section 2.3.4 discussing individual notice). Finally, the project developer should implement a mechanism to record that each external sharing process is conducted pursuant to a formal understanding, a Memorandum of

---

[22] DHS Secretary's Memorandum concerning the DHS Policy for Internal Information Exchange and Sharing, February 1, 2007.

Understanding or Agreement, or other comparable document, that details the nature and purpose of the external sharing.

3.3. Develop Dictionaries and Models

In order to ensure that each field of PII and each use of each field of PII is identified and justified, the project developer should create dictionaries that define the fields of PII and the processing of PII so that all of the teams involved in the project (technical, management, security, legal, privacy) can understand what and how PII will be used in the operational system. The project developer should also create models (data reference models and process flow diagrams) to visually present the interaction of the various data and use elements identified in the dictionaries so that all of the teams involved in the project can easily understand how the operational system will function once deployed. In addition to the contribution these dictionaries and models offer the review process, they also assist the developer verify the accuracy of all claims regarding what and how PII is to be collected and used in the system.

3.3.1. Develop a PII Data Dictionary

The project developer should create a dictionary that identifies and describes each field of PII to be used in the system. This dictionary should be written in plain language with a minimum of technical jargon or acronyms and should be concise and informative enough to serve as a single reference to evaluate the system from multiple perspectives (technical, management, security, legal, privacy). The PII data dictionary should include at least the following information regarding each field of PII:

- <u>Name</u>: The name of the data field – the technical name of the field of PII to be used in the system.

- <u>Description:</u> The description of the content of the field of PII – written for a non-technical audience.

- <u>Source</u>: The source of the field of PII – whether it derives from inside or outside the Federal Government, and if it originates outside the Federal Government, whether it comes from a commercial data provider.

- <u>Ownership</u>: The DHS Component governing how that field of PII may be used by the system – If there are multiple owners, then list all owners.

- <u>Justification</u>: The value justification for the field of PII including – a description of the necessity for the PII based on mission requirements and the authority to collect that field of PII.

3.3.2. Develop a Process Dictionary

As a companion to the data dictionary, the project developer should develop a process dictionary that describes each discrete process of the system and the important aspects of each process. Just as with the data dictionary, the process dictionary should be concise and informative so it can be used as a single reference to evaluate the system from multiple perspectives. The fields should include at least the following:

- Name: The name of the process.

- Description: The description of the process, written for a non-technical audience.

- Ownership: The process owner is the DHS component or external organization that is responsible for the successful operation of the process. There should be only one owner for the process.

- PII: The fields of PII from the data dictionary that are used by the process.

- Justification: The value justification for the process, including the authority to engage in the process.

3.3.3. Develop a Data Model

The project developer should create a data model that represents all data used by the system and specifically mark those data fields that contain PII - note that this means the scope of the data model will be broader than the PII data dictionary. The data model should identify each connection between those fields marked as PII and other data fields to show the full network of touch points to PII and verify accuracy of the PII data dictionary.

3.3.4. Develop a Process Flow Model

As a companion to the data model, the project developer should develop a process flow model (diagram) depicting all processes of the system and specifically mark those processes using PII. The process flow model should depict the relationships between each process and the data that is exchanged between processes. The process flow model should be general enough to describe the overall use of data (matched to the data dictionary) and to present the context for each use of each field of PII. Within the larger context of the overall use of information, the model should depict the uses of each field of PII with sufficient detail to verify the accuracy of the data dictionary and data model as well as all claims regarding what and how PII is used by the system.

3.4. Implement Data Quality Standards

The project developer should develop the project to meet the data quality standards developed by the DHS Data Management Working Group (DMWG). If no specific standards applies, the project developer should recommend a threshold for validating the accuracy, relevance, timeliness, and completeness, of each field of PII and coordinate with the DMWG to formally approve the project developer's approach.

Data quality should be defined in context. If the field of PII is migrated from one system to another or is otherwise used within a different environment, the data quality of the PII should be verified again within the new (receiving) context. It is possible that a field of PII could be deemed accurate in one context for one purpose and inaccurate in another context or as judged against another purpose.

The project developer should specifically examine each contemplated sharing of PII, internal and external to the Department, and evaluate the new context for each field of PII. Only when the field of PII can be determined to be accurate in fact (is the PII still true?) and in meaning (does the PII represent the same assertion?) should the field of PII be shared.

The project developer should build technical and procedural mechanisms into the project to evaluate the data quality of each field of PII and regularly review each field of PII to ensure that the applicable data quality standards are met.

3.5. Implement Individual Participation

The project developer should coordinate with the program manager regarding how notice, correction, and redress for the individual will be addressed and identify any functionality that must be implemented in the system in order to support the identified level of individual involvement. The below are general considerations to address.

3.5.1. Implement Notice upon Collection of PII

The project developer should develop a mechanism that describes, to the individual, the details and context regarding the collection of PII. This notice should be made available to the individual at the same time and presented through the same method the PII is collected. If PII is collected through an online system, the notice should be provided on the same screen. If PII is collected through a paper form, the notice should be printed on the form itself. The project developer should coordinate with the Component Privacy Officer or the Privacy Office for further details regarding this notice. The notice should generally explain:

- The authority enabling the collection of PII;
- The purpose for the collection of PII;
- Whether the collection of PII is mandatory[23] or voluntary;

---

[23] If the business rules state that the individual's consent is not needed, the project developer should document the authority supporting the mandatory submission of the field of PII.

- The effects of not providing the PII; and

- Whether the field of PII would be shared with third parties and if so, the identity of those third parties.

Implementing this type of notice is particularly important when collecting SSN.

3.5.2. Develop the Capability for Access and Correction

The project developer should develop mechanisms to enable the individual to access and correct each field of PII. This does not mean a user account should be provided to the individual for direct access to the data. This does mean that the project developer should develop the capability to support what could be an administrative process to update PII based on the individual's request. The system should be developed to:

- Identify PII related to an individual;

- Present those records in a way that would be meaningful to the individual (to enable review of each field of PII);

- Update or append each field of PII as requested; and

- Record that the field of PII was updated pursuant to the individual's request.

3.6. Schedule Data Retention and Removal

As soon as a field of PII is no longer necessary, the PII should be removed from use. The project developer should coordinate with the Office of the DHS Senior Records Officer to identify a data retention schedule for each field of PII in the system. This retention schedule must also meet NARA requirements.

If a field of PII is to be removed from the system and placed in an alternate data facility (e.g., data warehouse) for either archiving or retirement, the project developer should identify options to tier access to archived PII so that different standards apply for different categories of PII.[24]

The project developer should also develop the system to enable deletion of all fields of PII when the PII is scheduled or marked for removal. It is not sufficient to merely tag PII as "inactive" or "deleted." The PII must actually be removed.

---

[24] The project developer should coordinate with the Component Privacy Officer or the Privacy Office regarding how best to categorize the archived PII for purposes of privacy protection considerations.

3.7. Secure PII

The project developer should implement information security measures for each field of PII pursuant to DHS Management Directive 4300.1. Protection should be sufficient to prevent loss, unauthorized access, or unintended use of PII and should conform to all applicable Departmental information security requirements, including at least the following:

- Transmission of PII should occur through encrypted communication channels or other equivalently controlled mechanisms.

- Printing and copying PII should be prevented unless specifically authorized by the Designated Accrediting Authority and any documentation should be labeled according to DHS information security policies.

- Each individual user should have a uniquely identified account and each user account should be assigned a specific role within the system.

- A specific approval process should be implemented that governs the creation and updating of user accounts and each transaction related to the administration of accounts should be recorded including the account(s) associated with each transaction.

- The nature and use of all passwords should comply with DHS MD 4300.1 requirements.

The project developer should provide security categorizations of the technology as identified in NIST publication FIPS PUB 199, "Standards for Security Categorization of Federal Information and Information Systems." The FIPS 199 categories are: Confidentiality, Integrity, and Availability, and the ratings for each are Low, Moderate, and High. Systems using fields of PII must be designated with at least a moderate level of risk for confidentiality and all correlating controls should be implemented.

If the use of SSNs is authorized, the project developer must use the appropriate information security mechanism to limit access to SSNs to those with a valid need-to-know and must prevent posting of SSNs in any public areas including publicly accessible network folders. Additional security measures should be used to further limit access to SSNs such as automatic expiration of user access based on lack of use over a period of time and the use of encryption and password-based protection.

The project developer should identify the appropriate set of information security controls for the use of SSNs and obtain the required Privacy Office and OCISO approval of those controls before deploying SSNs in the system.[25]

Prior to transmitting PII externally, the project developer should confirm that the receiving system meets or exceeds the applicable DHS security standards. This determination should be recorded sufficiently to support a comprehensive compliance audit.

3.8. Log Activities

The project developer should develop the system to support a comprehensive audit of all collection and use of all fields of PII to ensure that the actual use of each field of PII aligns with system and privacy compliance documentation, including the functions that support individual involvement discussed in sections 2.3 (management) and 3.5 (deployment).

If the use of SSNs is authorized, the project developer must log access to SSNs and implement periodic reviews of the audit logs for compliance with the authorization.

Audit logs should not contain the actual content of fields of PII, but should include sufficient information to answer the following questions:

- What is the source of each field of PII?

- When was each field of PII accessed?

- What were the uses of each field of PII, when and by whom?

- When was each piece of PII last updated and why?

- Were there any suspicious transactions related to any field of PII and if so, what was suspicious and which users were involved?

If the system shares fields of PII internally (inside DHS), the project developer should design and deploy the logs to record sufficient information to support an audit of whether each field of PII was shared pursuant to a determination that the recipients needed the field of PII to successfully perform their duties, possessed the requisite security clearance, and provided assurance of appropriate safeguarding and protection of the PII. To the extent that the full breadth of this determination may be beyond the scope of what can be logged about the system's operation, the project developer should coordinate with the program manager to identify other administrative mechanisms that could support a privacy compliance audit.

---

[25] The project developer should also obtain approval from the Component Privacy Office and Component Information System Security Manager.

If the system shares fields of PII externally (outside DHS), the project developer should develop technical and procedural mechanisms to record the fact of each transmission of PII with sufficient detail to support providing notice to individuals, (see sections 2.3.4 and 3.2.2 - discussing the capability to respond to individual requests) and an audit of the actual external sharing of information compared to privacy compliance documentation.

4. Summary

The privacy compliance documentation (PTAs, PIAs, and SORNs) embody the collaboration of program, technical, legal, security, and privacy teams across the Department. PIAs and SORNs are published to the public in order to foster transparency and individual participation regarding how DHS uses PII to fulfill its mission.

The PTIG is designed to assist technology managers and developers more easily identify the full set of individual privacy protections found separately in the various components and policies of the DHS privacy compliance process. The privacy considerations in the PTIG, organized for IT managers and developers, should enable further integration of privacy protections into the development life cycle thus improving the efficiency and effectiveness of the development of privacy protective technology and documentation. Appended to this guide is a short checklist that summarizes the primary privacy protection considerations for both managers and developers.

For more information regarding this guide and the privacy compliance requirements and process, please contact the Privacy Office via phone: 703-235-0780, email: privacy@dhs.gov, or visit the Privacy Office's website: www.dhs.gov/privacy.

5. APPENDIX: Privacy Compliance Requirements Checklists

The following checklists distill the PTIG into a concise set of specific privacy considerations that should be incorporated into the management and development of operational IT systems to enhance privacy protections. As noted above, the recommendations in the PTIG are not requirements. The only mandate is the existing privacy compliance analysis and documentation requirements embodied in the PTAs, PIAs, and SORNs.[26] This checklist, like the guide itself, is organized by role: program manager and project developer.

---

[26] As noted throughout the PTIG, There may be other obligations or exemptions from obligations that apply to the development and management of a particular IT system. IT managers and developers are strongly encouraged to contact the Privacy Office and OGC to address the specific applicable requirements.

PROGRAM MANAGER

The following are specific privacy protections for the program manager to consider:

- ☐ Ensure the Fair Information Practice Principles (FIPPs) guide the managed collection and use of PII:
    - ☐ Transparency
    - ☐ Individual Participation
    - ☐ Purpose Specification
    - ☐ Data Minimization
    - ☐ Use Limitation
    - ☐ Data Quality and Integrity
    - ☐ Security
    - ☐ Accountability and Auditing
- ☐ Identify information required to generally describe the project
- ☐ Identify any PII related to the project and/or used in the system(s)
- ☐ Ensure Controlled Operations, including:
    - ☐ Itemized review of PII for relevance, necessity, and contextual integrity
    - ☐ Structured development life cycle
    - ☐ Formal change management process
    - ☐ Alignment of business drivers and measures of success
- ☐ Ensure individual participation, including:
    - ☐ Direct collection of PII from individual
    - ☐ Individual is provided access, notice, correction, and redress
- ☐ Coordinate all Documentation
- ☐ Complete Privacy Compliance Documentation
    - ☐ Privacy Threshold Analysis
    - ☐ Privacy Impact Assessment
    - ☐ System of Records Notice
- ☐ Provide training
- ☐ Enable auditing of the use of PII

PROJECT DEVELOPER

The following are specific privacy protections for the project developer to consider:

- ☐ Ensure the Fair Information Practice Principles (FIPPs) guide the managed collection and use of PII:
    - ☐ Transparency
    - ☐ Individual Participation
    - ☐ Purpose Specification
    - ☐ Data Minimization
    - ☐ Use Limitation
    - ☐ Data Quality and Integrity
    - ☐ Security
    - ☐ Accountability and Auditing
- ☐ Minimize Collection of PII
- ☐ Limit use of PII
- ☐ Develop Dictionaries and Models
- ☐ Implement Data Quality Standards
- ☐ Schedule Data Retention and Removal
- ☐ Secure PII
- ☐ Log Activities